THE WORLD'S
VANISHING ANIMALS
The Mammals

the world's
VANISHING ANIMALS

written by **CYRIL LITTLEWOOD**
director of the Youth Service
of the World Wildlife Fund

illustrated by **D.W.OVENDEN** *F.Z.S*

THE MAMMALS

W. FOULSHAM & Co. Ltd.
LONDON · NEW YORK · TORONTO
SYDNEY · CAPE TOWN

W. FOULSHAM & CO., LTD.,
Yeovil Road, Slough, Bucks., England.

Printed in the Republic of Ireland by Cahill & Co., Limited, Parkgate Printing Works, Dublin 8.

CONTENTS

INTRODUCTION

This book is intended for young people of all ages who are in any way interested in the country-side and its wildlife, the wildlife of the world in general and those who feel that we must do everything possible to save the wildlife and wild places of the world now, before it is too late, so that future generations may be able to inherit them from us.

You see, during the past seventy or eighty years, the wild places of the world have been disappearing at an ever-increasing rate as Man has invaded them and taken them over for his own use. Forests, jungles and woodlands have been torn down, marshlands have been drained, lush valleys have been drowned beneath the waters of hydro-electric projects and the rivers and streams have been polluted by rubbish or poisonous chemical waste from factories. The wildlife living in these places has either fled or been destroyed . . . and such is Man's demand for land that the wildlife and wild places of the world are shrinking away faster and faster each year!

Unless something is done to check this unnecessary and wanton destruction now it will soon be too late and the wildlife of the world as we know it today will be lost to us within the span of only a few years.

Since 1900 alone, about one hundred species of wild animals have been wiped out by Man. Today there are about one thousand species facing the threat of extinction and many of these are already beyond any hope of saving.

This book deals with just sixty-nine different species of mammals which may soon disappear from the world. You will see that we have illustrated them all and that beside their names we have placed one of several symbols which will also be found on the maps of the world. I have used these marks to show areas where these vanishing animals still live, but it must be remembered that they are not necessarily confined within the area of the symbols. Among these animals you will find many that you have known since you were very young; animals like the Indian Lion, the Polar Bear, the Panda, Cheetah, Wild Horse, Otter, Gorilla, Orang-utan and the Koala. Surely these animals are all worth preserving for the benefit of future generations?

To encourage further volumes in this series dealing with birds and other endangered species it would help me considerably if I knew just how many young people (and adults) have found this book interesting and so I am providing the address of my office just in case you would like to write to me.

Having survived millions of years of evolution, birds are vanishing from the sky, fish from the sea and other creatures from the land . . . and the threat to wildlife is now very grave indeed. I hope that every young person and adult who reads this book will feel sufficiently interested in the preservation of wildlife to want to do something to help for without this interest and support it seems certain that all wildlife is doomed to vanish before long.

Address:
Cyril Littlewood,
WILDLIFE,
Morden, Surrey.

Founder and Director,
The Wildlife Youth Service.

AFRICA & ASIA MINOR

The continent of Africa lies on both sides of the equator and covers an area of about 11,500,000 square miles. The African coastline measures approximately 18,950 miles in length and its shores are washed by the Atlantic Ocean, Indian Ocean, Red Sea and Mediterranean Sea.

The zones of Africa are well marked. In the north-west there is a narrow strip of sub-tropical grassland with wooded mountains. Just to the south of this lies the desert region of the Sahara which stretches from the Red Sea to the Atlantic. Beyond the desert and moving south again the terrain passes into tropical grassland. To the east of this region there are open forests, park-like areas, swamps and highlands; and to the west the great tropical forests from Liberia to the Congo. The lower third of the continent consists of open plains, parkland, forests and the desert regions of the south-west.

Each zone has its own specialized forms of wildlife, which although abundant in the past have now become rare. Careful protection will be essential if the wildlife of Africa is to survive in the future.

MOUNTAIN GORILLA
(Gorilla gorilla beringei) △

Order: *Primates*
Family: *Pongidae*

The adult male Mountain Gorilla is the largest and most powerful of all primates and yet in spite of its immense strength and ferocious appearance it is a shy and generally inoffensive creature unless provoked.

These huge apes live in the mountainous highlands of the eastern Congo and south-western Uganda, where they move through the forests in small family groups feeding by day and sleeping in the trees at night.

It has been estimated that between 5,000 and 15,000 of these animals still survive in Africa.

The Lowland Gorilla a closely related race lives in west Africa.

Description: Heavily built, tailless ape with powerful arms and weak legs. The old males have a prominent grey or silver 'saddle' across the back but are otherwise covered with long, shaggy, black hair. Males also develop a high crown to the skull and a pronounced ridge above the eyes. The eyes are small, the nostrils large, and the face generally hairless. Teeth are large and strong.

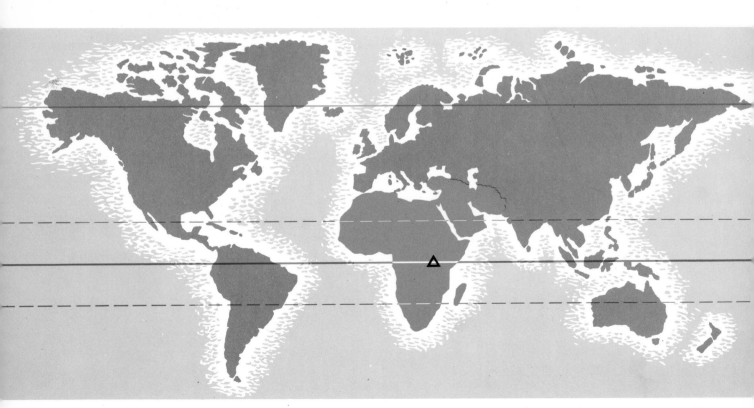

MOUNTAIN GORILLA
(*Gorilla gorilla beringei*).
Adult males may reach a height of 6 ft. (1·83 metres), and weigh up to 450 lbs. (204 kgs.). Record weight for male
Mountain Gorilla 716½ lbs. (244·9 kgs.).   *Food:* Vegetables and fruits.

ARABIAN ORYX
(*Oryx leucoryx*) △

Order: *Artiodactyla*
Family: *Bovidae*

Of the three surviving species of Oryx this is the smallest. Once found throughout Arabia and in Syria and Mesopotamia, the Arabian Oryx has been brought to the very brink of extinction through uncontrolled hunting and it has been estimated that fewer than 500 remain in the wild. When seen from a distance this animal looks a little like a white horse with a single horn growing from its forehead—and so it was the Oryx that gave us the legend of the unicorn. They inhabit gravelly desert country and can go without water for long periods. Thanks to the success of an expedition in 1962, two male and one female Oryx were captured and eventually sent to a zoo in Phoenix, Arizona to form the nucleus of a breeding herd. Two more small captive herds were established in Arabia and now the Arabian Oryx (in captivity at any rate) is safe.

Description: Almost pure white in colour, with dark chocolate markings on the legs. Slight shoulder hump. Both sexes carry horns. Tail white with prominent black tuft. Black markings on face.

PERSIAN FALLOW DEER
(*Dama mesopotamica*) ☐

Order: *Artiodactyla*
Family: *Cervidae*

Although generally similar to the European Fallow Deer in appearance, the rare Persian species does display certain differences. Persian Fallow Deer are larger than those found in Europe and they are much more brightly coloured. The antlers of the European species are always flattened (palmated) at the tips—those of the Persian Fallow Deer are not. The species was thought to be extinct until 1875, when it was rediscovered by Sir Victor Brooke. In 1917 it was once again declared to be extinct. As recently as 1955, however, the species was discovered living in dense riverine vegetation in south-western Iran. It has been estimated that only about 50 of these animals remain in the wild state.

Description: Rich reddish-fawn coat marked with white spots on the neck and upper parts of the body and white dorsal stripes.

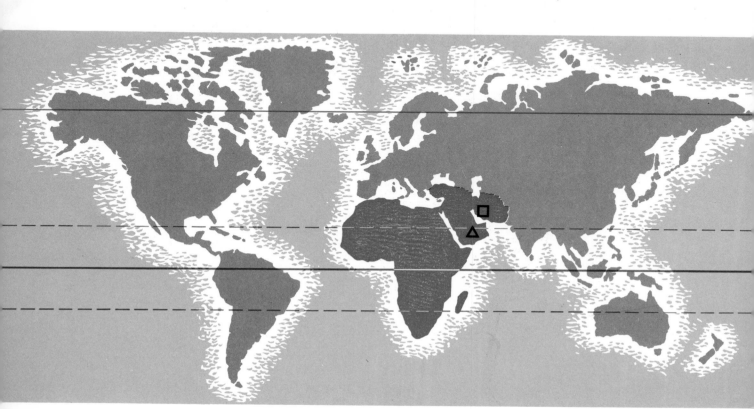

ARABIAN ORYX
(*Oryx leucoryx*).
Shoulder height 3½ ft. (1·07 metres).
Weight about 300 lbs. (136 kgs.).
Horns straight and up to 29 inches
(0·73 metres) in length. *Food:* Desert
vegetation.

PERSIAN FALLOW DEER
(*Dama mesopotamica*).
Shoulder height about 3½ ft. (1·07 metres). Antlers
36–40 inches (0·91–1·01 metres) in length. *Food:*
Grasses and leaves.

SCIMITAR-HORNED ORYX
(*Oryx tao*) △

Order: *Artiodactyla*   Family: *Bovidae*

The Scimitar-Horned Oryx now inhabits a narrow strip of land south of the Sahara, on the fringes of the desert. In spite of the drastic reduction in its range and numbers this oryx is still comparatively plentiful. Constantly hunted, this animal will need careful protection in the future if it is to escape the fate of the Arabian species. These antelopes move in large herds.

Description: Whitish in colour with a faint chestnut tinge on face, neck, flanks and upper limbs. The horns curve backwards. Both sexes carry horns. Slight hump on shoulder.

GIANT ELAND
(*Taurotragus derbianus*) ☐

Order: *Artiodactyla*   Family: *Bovidae*

This is the largest living antelope. Only a few dozen of these animals are thought to exist. The surviving Giant Eland live in three isolated areas on the borders of Senegal, Mali and Guinea, in West Africa. Eland move in small herds of a dozen or so. In spite of their bulky build they can make surprisingly high jumps of nearly 6 ft. (1.83 metres) in order to clear obstacles.

Description: A heavily built antelope with a stout body and humped shoulders. The coat is reddish-brown in colour with about 14 vertical white stripes on the sides. There is a well developed dewlap under the throat. The horns are spirally twisted and straight. Both sexes carry horns.

ADDAX
(*Addax nasomaculatus*) ○

Order: *Artiodactyla*   Family: *Bovidae*

The Addax has become adapted to life in the desert—getting its moisture only from the plants on which it feeds. There are thought to be about 5,000 Addax left in the wild.

Description: Sandy coloured with white underparts. Tuft of chestnut-coloured hair on forehead. Coat colour varies seasonally and is darker in the winter. Both sexes carry spirally twisted horns.

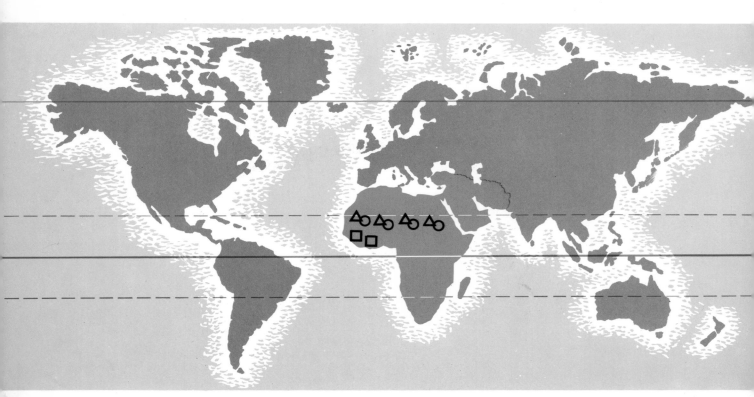

SCIMITAR-HORNED ORYX
(*Oryx tao*).
Shoulder height 4 ft. (1·22 metres).
Weight about 450 lbs. (204 kgs.).
Horns up to 45 inches (1·14 metres) in length. *Food:* Seasonal grasses, leaves and desert vegetation.

GIANT ELAND
(*Taurotragus derbianus*).
Shoulder height about 6 ft. (1·83 metres). Weight up to 2,000 lbs. (907 kgs.). Horns up to 42 inches (1·07 metres) in length. *Food:* Chiefly leaves—but will also graze.

ADDAX
(*Addax nasomaculatus*).
Shoulder height 3½ ft. (1·07 metres).
Horns up to 43 inches (1·08 metres) in length. *Food:* Desert vegetation.

MOUNTAIN NYALA
(*Tragelaphus buxtoni*) △

Order: *Artiodactyla*
Family: *Bovidae*

This very handsome short-haired antelope is the largest member of the Bushbuck tribe and it lives in the mountains of southern Ethiopia. The Mountain Nyala was discovered as recently as 1908 by Major Ivor Buxton (you will notice that he receives credit for this discovery in the scientific name for this animal). This discovery caused great excitement among naturalists—and, unfortunately, the sportsmen of that time. For the naturalists, a new and interesting animal had been found at a time when it was felt that all Africa's big animals had been catalogued. For the sportsmen—here was a magnificent new trophy to hunt down in difficult and dangerous country. One of the most interesting points about the discovery of the Mountain Nyala was that it lives some 2,500 miles distant from its near (but smaller) relative the Nyala (*Tragelaphus angasi*) of south-east Africa. Mountain Nyalas live at altitudes ranging from 9,000 (2,743 metres) to over 11,500 ft. (3,505 metres) and in zones of vegetation ranging from forest to high, hot heathland. During the day the sun scorches down on the forests and heaths—but at night the temperature falls to near-freezing point, so this antelope has to contend with very different conditions by day and night. In spite of the interest shown by naturalists and hunters in this antelope—little is known about its life and habits. Until recently it was feared that only about 2,000 Mountain Nyala were left in the wild. However, an expedition organised jointly by the World Wildlife Fund and the American National Geographic Society made a survey of the species in 1966 and came to the conclusion that at least 4,500—and perhaps up to 10,000 of them still survive in the mountains of Ethiopia. Unfortunately, the Mountain Nyala is not well protected—and a great many are shot by illegal hunters. This antelope will need carefully protecting if it is to survive.

Description: Large antelope, greyish-chestnut in colour with white markings on the chin, throat and chest. There are also white vertical stripes on the back, some white spots on the flanks and a black and white dorsal crest. The underparts, underside of tail and inside of legs are also whitish. The horns (males only) are blackish with two spiral twists.

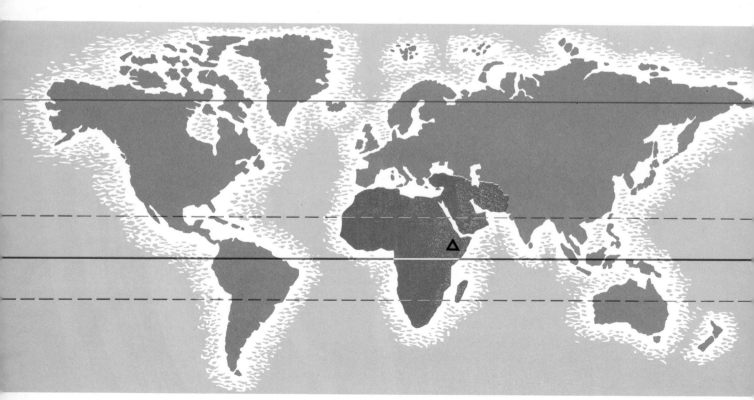

MOUNTAIN NYALA
(*Tragelaphus buxtoni*).
Shoulder height 4½ ft. (1·37 metres). Weight about 450 lbs. (204 kgs.). Horns 42-44 inches (1·07-1·11 metres) long.
Food: Grasses and leaves.

DORCAS GAZELLE
(*Gazella dorcas*) △

Order: *Artiodactyla* Family: *Bovidae*

One of the smallest of Gazelles, the Dorcas lives in parts of North Africa and Saudi Arabia. These gazelles inhabit inhospitable plains and desert regions—and can survive for long periods without water just as long as they can find vegetation to eat.

Description: Rather smaller than a roe deer, with reddish-brown hair on the upper parts and flanks and a sandy-whitish colour below. The two colours are separated by an indistinct dark stripe along the flanks. These colours vary slightly between the three recognised races.

GIANT SABLE ANTELOPE
(*Hippotragus niger variani*) ☐

Order: *Artiodactyla* Family: *Bovidae*

This strikingly handsome antelope is the largest of the four races of Sable—and the last of the large animals of Africa to be discovered (in 1913). The total population of Giant Sable left in the wild is estimated to be between 500–700. These animals occupy the Luando Reserve and the Cangandala area in Angola, West Africa. There are no specimens in captivity.

Description: Adult bulls are jet black in colour, with white underparts and face markings and well developed black manes. The females are lighter in colour with smaller horns. The horns of the males are long and sweeping—and they have caused this antelope to be much sought after as a hunting trophy.

WHITE-TAILED GNU or Black Wildebeest
(*Connochaetes gnou*) ○

Order: *Artiodactyla* Family: *Bovidae*

There are two species of Gnu, the other one being the common Brindled Gnu (or Wildebeest). The White-tailed species is limited to parts of South Africa—where the population numbers some 2,000 specimens in game reserves.

Description: Dark brown in colour—but dark enough to appear black. There is a mane of stiff hairs coloured black and off-white. A beard extends from the lower jaw to the chest and the whitish horse-like tail reaches almost to the ground. Both sexes have horns.

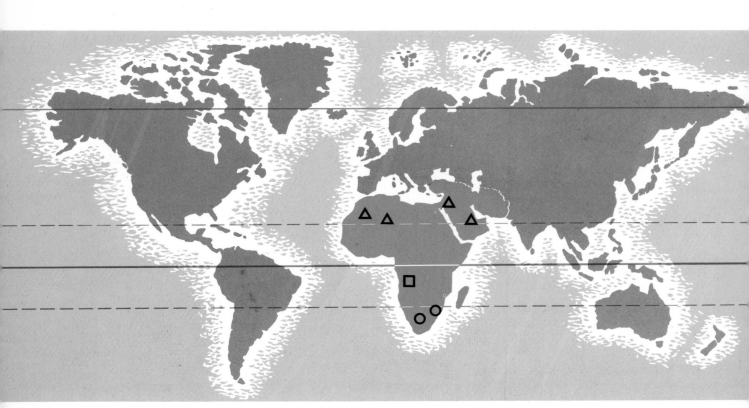

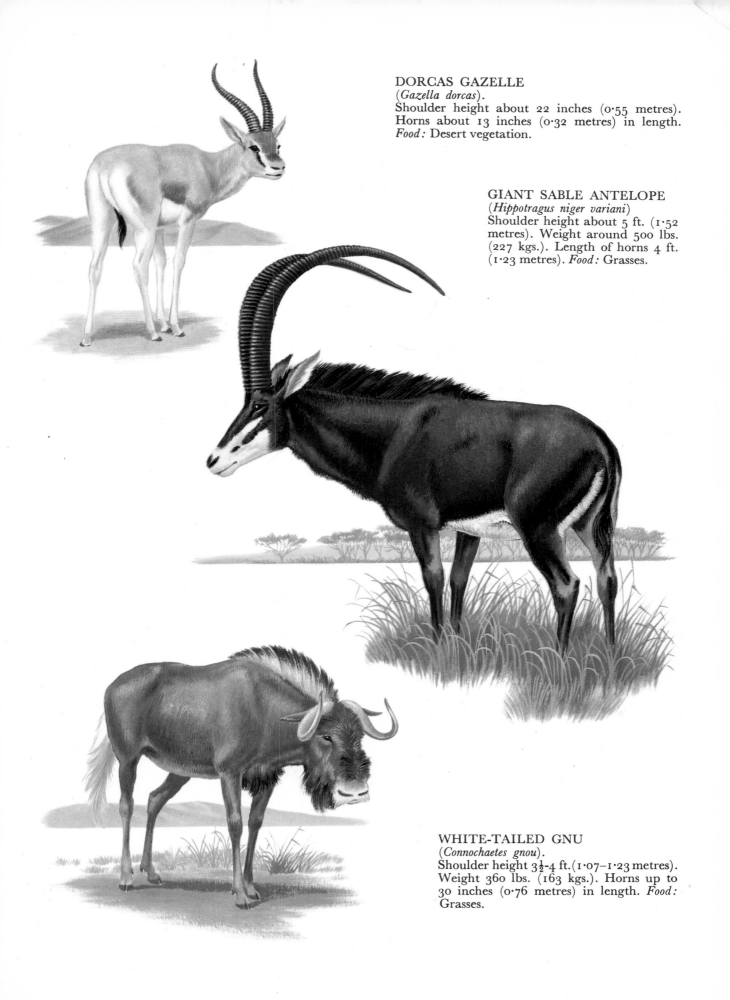

DORCAS GAZELLE
(*Gazella dorcas*).
Shoulder height about 22 inches (0·55 metres).
Horns about 13 inches (0·32 metres) in length.
Food: Desert vegetation.

GIANT SABLE ANTELOPE
(*Hippotragus niger variani*)
Shoulder height about 5 ft. (1·52
metres). Weight around 500 lbs.
(227 kgs.). Length of horns 4 ft.
(1·23 metres). *Food:* Grasses.

WHITE-TAILED GNU
(*Connochaetes gnou*).
Shoulder height 3½-4 ft.(1·07–1·23 metres).
Weight 360 lbs. (163 kgs.). Horns up to
30 inches (0·76 metres) in length. *Food:*
Grasses.

BROWN HYAENA
(Hyaena Brunnea) △

Order: *Carnivora*
Family: *Hyaenidae*

This South African species of hyaena is found mainly in south-west Africa and south-western Bechuanaland. It is also known to occur in the Kruger National Park and has been recorded in Portuguese East Africa and Southern Rhodesia. Although the hyaenas tend to resemble large, ungainly dogs—they are classified in a family of their own and are more closely related to cats than dogs. One of Nature's scavengers, the Hyaena feeds mainly on carrion and performs a useful service by helping to dispose of the remains of animals killed by other carnivores. In South Africa this hyaena is also known as the Strandwolf. Population in the wild is not known.

Description: Smaller and more lightly built than its close (and more common relative) the Striped Hyaena, this animal has the typical massive head and powerful jaws common to the family. The ears are large, thin and pig-like. The hair on the back and sides is very long—and may measure up to 10 inches (0.25 metres) in length. There is no mane or crest. The powerful jaws and strong

teeth are capable of crushing the thigh bones of a buffalo! Nocturnal by nature.

CAPE MOUNTAIN ZEBRA
(Equus zebra) ☐

Order: *Perissodactyla*
Family: *Equidae*

There are two races of Mountain Zebra, the Cape (illustrated) and Hartmann's (Equus zebra hartmannae) of which the latter is the most numerous. About 80 Cape Mountain Zebras still survive in South Africa's Mountain Zebra National Park (Cape Province) and the species now seems fairly safe. This is the smallest of the zebras and it is quite donkey-like in general appearance.

Description: The stripes are narrow and close-set, except on the rump, where they widen and broaden. A unique 'grid-iron' stripe pattern on the rump helps to distinguish this zebra from other species. The ears are long and the hooves narrow. The striping extends down the legs to the black hooves but there are no stripes on belly.

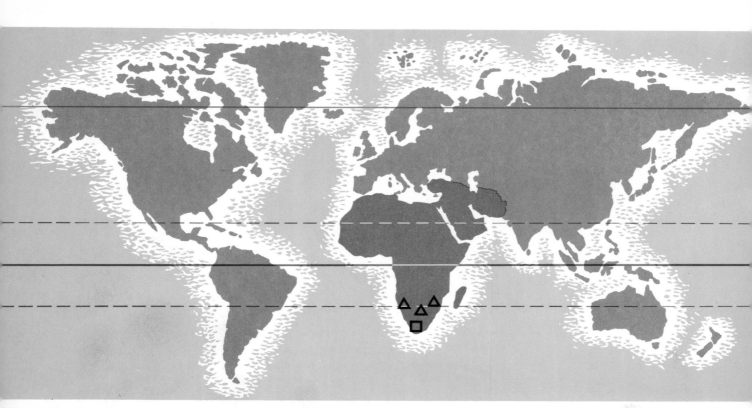

BROWN HYAENA
(*Hyaena brunnea*).
About 3 ft. (0·91 metres) in length. Tail
12-14 inches (0·30–0·35 metres). Shoulder
height about 25 inches (0·63 metres).
Weight around 60-70 lbs. (27–32 kgs.).
Food: Chiefly carrion.

CAPE MOUNTAIN ZEBRA
(*Equus zebra*).
Stands 4 ft. (1·22 metres) at the shoulder. Length
of life 25 years. *Food:* Grass.

BLACK RHINOCEROS
(Diceros bicornis) △

Order: *Perissodactyla* Family: *Rhinocerotidae*

The Black (or Hook-lipped) Rhinoceros is the most numerous of all the five species surviving in the world today and about 13,000 still exist in eastern and southern Africa. Unlike the White Rhinoceros this animal is a browser and has a pointed and prehensile upper lip. It is solitary by nature and quite aggressive. This species inhabits thick bush-country and is distributed over an extensive range chiefly in eastern and central Africa.

Description: Rather smaller than the White Rhinoceros with hooked upper lip. Has no incisor teeth. Skin slightly darker grey. Hide thick and almost hairless, except for a small tail tuft. Two horns—composed of tightly compressed fibres (not ivory or bone). The front horn is the larger. Three toes on each foot.

WHITE RHINOCEROS
(Diceros simus) ☐

Order: *Perissodactyla* Family: *Rhinocerotidae*

The White (or Square-lipped) Rhinoceros is not only the largest member of its family but also the second largest land animal. It now numbers about 3,900 in the wild and this population inhabits the grassy open plains in Zululand, Uganda, south-western Sudan and parts of the Congo.

Description: Solitary and inoffensive by nature. Skin slate-grey in colour and hairless except for a fringe on the edge of the ear. There is also a tuft of hair on the tail. Front horn long and slender— rear horn conical, short and straight. Massive head. Hump on shoulder. Eyesight poor. Acute sense of smell.

PYGMY HIPPOPOTAMUS
(Choeropsis liberiensis) ○

Order: *Artiodactyla* Family: *Hippopotamidae*

Generally similar in appearance to its larger relative—the Pygmy Hippopotamus has a smaller head, longer legs and shorter body. The name Hippopotamus is derived from two Greek words meaning 'river horse'—although in actual fact the Hippo is more closely related to the pig. This small hippo is found in Liberia, Sierra Leone and part of southern Nigeria, and moves in pairs. It is less aquatic than the larger hippopotamus.

Description: A stoutly built animal about the size of a pig. Nostrils are valve-like slits and widely separated. The small external ear can be folded when submerged. It has only one pair of lower incisors (the larger hippo has two pairs).

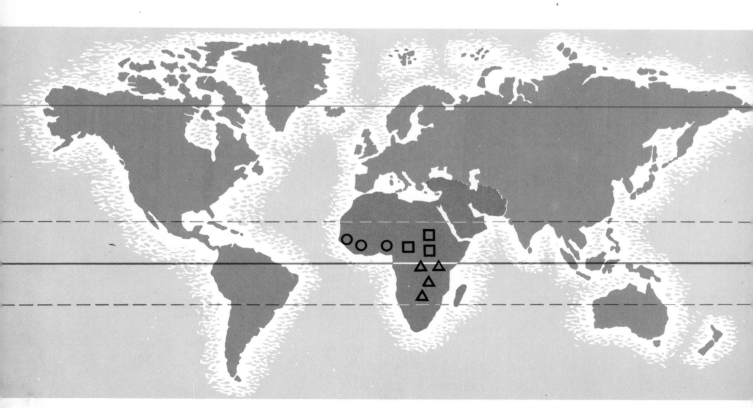

BLACK RHINOCEROS
(*Diceros bicornis*).
Length about 11-12 ft. (3·35-3·66 metres). Shoulder height 5½ ft. (1·68 metres). Weight up to 2 tons (2,032 kgs.). Calf 75 lbs. (34 kgs.) at birth. Front horn up to 53½ inches (1·31 metres) in length. Rear horn up to 20 inches (0·49 metres). *Food:* Leaves, twigs and herbage.

WHITE RHINOCEROS
(*Diceros simus*).
Length 14½ feet (4·42 metres). Shoulder height 6½ ft. (1·98 metres). Weight up to 4 tons (4,064 kgs.). Front horn up to 62½ inches (1·61 metres) in length. Rear horn up to 21 inches (0·51 metres). *Food:* Grass.

PYGMY HIPPOPOTAMUS
(*Choeropsis liberiensis*).
Length 5-6 ft. (1·52-1·83 metres). Tail 7 inches (0·17 metres). Shoulder height 2½ ft. (0·76 metres). Weight 400-500 lbs. (181-227 kgs.). *Food:* Herbage—especially grass.

CHEETAH
(*Acinonyx jubatus*) △

Order: *Carnivora* Family: *Felidae*

The Cheetah is the fastest animal on land. This cat can attain speeds of 60 m.p.h. and over. The Asian race has become almost extinct. In Africa too, the Cheetah has gone from many of its former haunts and seems to be declining in numbers. The Cheetah inhabits open grassland and semi-desert regions south of the Sahara and it can still be found in parts of Sudan, Senegal, Bechuanaland, Transvaal, Rhodesia and East Africa.

Description: Slender build, long legs and small head. Unlike other cats, the Cheetah is unable to retract its claws. The short coat is yellowish in colour, with dark blackish spots which cover the body, legs and upper part of the tail. It has small rounded ears and a black face stripe extending from eye corner to mouth. Whitish underparts. Usually solitary. Diurnal.

LEOPARD
(*Panthera pardus*) ☐

Order: *Carnivora* Family: *Felidae*

Seriously threatened in every part of its range, Leopards are heavily hunted for their skins or because of the threat they bring to human and domestic animal life. Among the races now facing extinction are the following: Clouded Leopard (S.E. Asia), Formosan Leopard (Taiwan), Korean Leopard (N.E. Asia), Transcaucasian Leopard (S.W. Asia), Sinai Leopard (Sinai) and Arabian Leopard (Arabia). In Africa, the Barbary Leopard (Morocco and Algeria) is down to the last 150 or so survivors.

Description: Size and colour may vary according to habitat. A large, well-built cat with a long tail and vicious nature. The coat colour and length of fur varies. Melanistic (all black) leopards are not uncommon—especially in rain forests. The 'Black Panther' is simply a melanistic leopard and not a separate species. Solitary and nocturnal.

OKAPI
(*Okapia johnstoni*) ○

Order: *Artiodactyla* Family: *Giraffidae*

Although not on the danger list this rare and interesting animal is well worth our attention.

Unknown to science until about 1900, the Okapi is the only living relative of the giraffe and inhabits the tropical rain forests of the Congo.

Description: The body is short and compact, the neck and legs comparatively long. Horns (2) in males only. Females are larger than males. Shy and nocturnal.

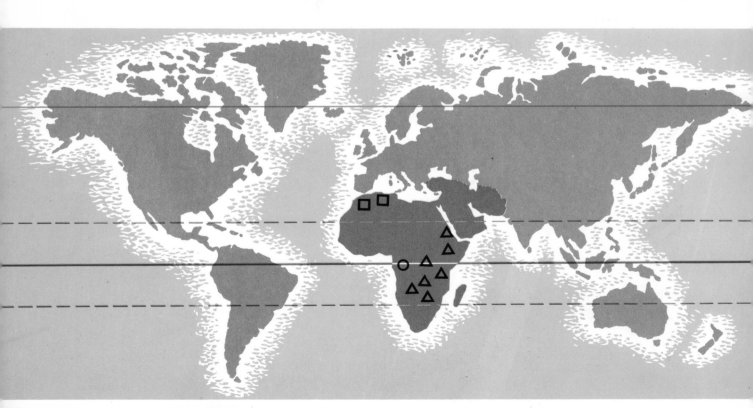

CHEETAH
(*Acinonyx jubatus*).
Length of head and body 5 ft. (1·52 metres). Tail 2½ ft. (0·76 metres). Shoulder height 2½ ft. (0·76 metres). Weight 100 lbs. (45 kgs.). *Food:* Gazelles, antelopes, ostriches and other medium sized and small mammals and birds.

LEOPARD
(*Felis pardus*).
Length of head and body 4½ ft. (1·37 metres). Tail 3 ft. (0·91 metres). Shoulder height 2 ft. (0·60 metres). Weight up to 150 lbs. (68 kgs.). *Food:* A wide variety of animals ranging from antelopes and domestic livestock to monkeys, porcupines and dogs.

OKAPI
(*Okapia johnstoni*)
Shoulder height about 5 ft. (1·52 metres).
Food: Leaves.

BLACK LEMUR
(Lemur macaco macaco) △

Order: *Primates*   Family: *Lemuridae*

True lemurs are found only in Madagascar. There are about 20 species. The Black Lemur is now confined to the north-west of Madagascar and two offshore islands. The actual number of these lemurs still surviving is not known.

Description: Males are black, females are brown to reddish-brown, with pale ear and cheek tufts. Both sexes have well developed ear-fringes.

SIFAKA
(Propithecus verreauxi) △

Order: *Primates*   Family: *Indriidae*

The Sifaka is a monkey-like lemur with long, silky fur, long hind-legs and a long tail. It lives in groups of 6-8 individuals. Diurnal.

Description: Coat mainly whitish-yellow in colour. The top of the head is black or blackish-brown with a white band on forehead. White cheek-fringe. Its face is hairless, and the palms of hands and soles of feet are black-skinned.

INDRI
(Indri indri) △

Order: *Primates*   Family: *Indriidae*

Inhabiting the mountain forests of eastern Madagascar, the Indri is the largest of all the lemur family, but is without the long tail. It moves in small, friendly troops of 5-6. Diurnal.

Description: Fur dense and silky, usually black and white, but very variable. Face, hands and feet black. Long muzzle, tufted ears, long hind legs and stump-tail.

AYE-AYE
(Daubentonia madagascariensis) △

Order: *Primates*   Family: *Daubentoniidae*

The Aye-aye is yet another strange animal from the forests of Madagascar. This little animal is particularly rare and probably fewer than 50 of them survive.

Description: Coarse brown hair and long bushy tail. Long hands with spidery clawed fingers. Rodent-like chisel shaped incisor teeth and no canine teeth. Large eyes.

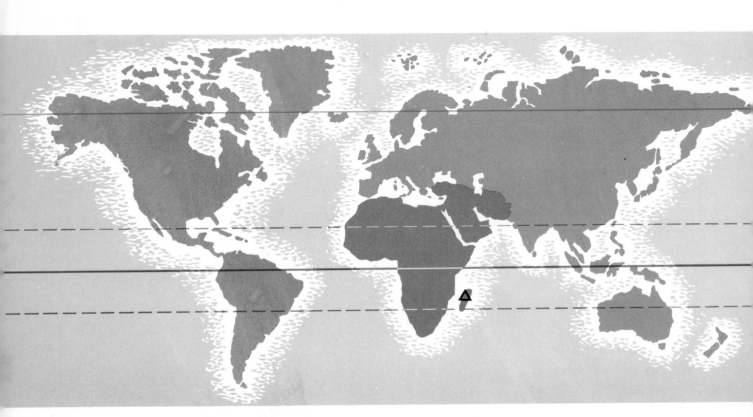

BLACK LEMUR
(*Lemur macaco macaco*).
Length of body 1½-2 ft. (0·46-0·60 metres).
Tail 1½ ft. (0·46 metres). *Food:* Fruits, small
animals, birds eggs, etc.

SIFAKA
(*Propithecus verreauxi*).
Length of head and body 18 inches (0·46
metres). Tail 22 inches (0·55 metres). *Food:*
Chiefly leaves and bark.

INDRI
(*Indri indri*).
Length 3 ft. (0·91 metres). Tail
1–2 inches (2·45–4·9 cm.). *Food:*
Chiefly leaves.

AYE-AYE
(*Daubentonia madagascariensis*).
Length of body 1½ ft. (0·46 metres). Tail
16 inches (0·40 metres). *Food:* Fruit and
insects.

THE AMERICAS

The two great continents of North and South America are joined together by the narrow land-link formed by Central America. The Americas got their name from the early explorer Amerigo Vespucci. North and Central America cover an area of about 9,354,570 square miles. The coastline measures about 47,000 miles in length. The longest river is the Mississippi, which with the Missouri and three minor rivers also included, have a watercourse some 3,710 miles in length. The highest point in North and Central America is Mt. McKinley (Alaska)—20,320 ft. (6,194 metres).

South America covers an area of 6,982,000 square miles. The coastline is 17,800 miles in length. The longest river is the Amazon—with a length of about 3,900 miles. This is also the largest river (in volume) in the world, although the Nile is the longest. The highest point is Mt. Aconcagua (Argentina) with a height of 22,834 ft. (6,962 metres). Stretching as it does from the Arctic Circle down towards the South Pole, the New World covers an enormous area and has a very wide range of climate and terrain. The wildlife of this great region is equally varied and well worth preserving for the future.

SPECTACLED BEAR
(*Tremarctos ornatus*)   △

Order: *Carnivora*   Family: *Ursidae*

Also known as the Andean Black Bear, the Spectacled Bear is the only member of the family Ursidae found in South America. It is now confined to a limited range which will become smaller and smaller as the remaining Andean forests are encroached upon by human progress. At present, however, it would seem that the Spectacled Bear is largely confined to Ecuador and northern Peru. Little is known about the habits of these bears in the wild, but they seem to thrive well in zoos. It seems generally accepted that this is the least carnivorous of all bears. No estimate has been made of the surviving population in the wild but about 50 Spectacled Bears are living in world zoos at the present time.

Description: This is a small bear with a shaggy, but thin black coat. The muzzle hair is buff coloured and this colour also encircles the eyes . . . hence the name 'Spectacled'. There is also a patch of buff coloured hair on the chest. These animals are nimble climbers and they are diurnal by nature.

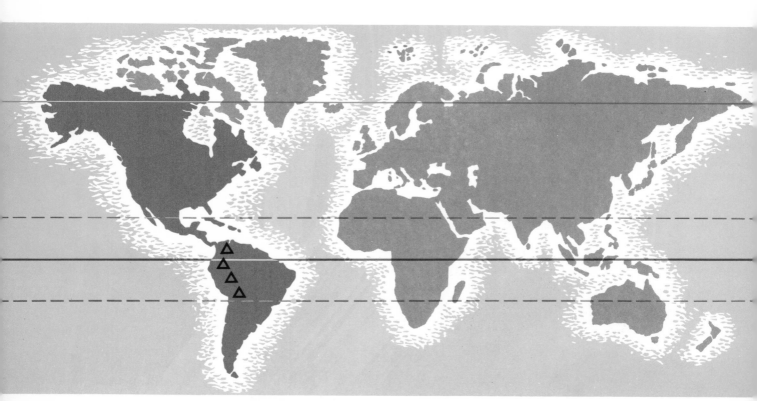

SPECTACLED BEAR
(*Tremarctos ornatus*).
Length of body 3½ ft. (1·07 metres). Shoulder
height 2-2½ ft. (0·61-0·76 metres). *Food:*
Leaves, fruit, nuts and some flesh.

GOLDEN LION MARMOSET
(*Leontideus rosalia*) △

Order: *Primates*   Family: *Callithricidae*

This family of small South American near-monkeys numbers 33 species. They are forest dwellers and, in large troops, inhabit the tree tops of south-eastern Brazil.

Description: A glossy, golden-yellow coat, with lion-like mane around head and shoulders. The fur is soft and silky. Long tail. Round head and round eyes. Face and ears hairless. Shy and delicate. Diurnal.

MANED WOLF
(*Chrysocyon brachyurus*) ☐

Order: *Carnivora*   Family: *Canidae*

Apart from the true Wolves of the Northern regions, the Maned Wolf is the largest member of the dog family and certainly one of the most striking. This animal lives in parts of Brazil, eastern Bolivia, Paraguay and north-eastern Argentina. Strictly speaking, it is a wild dog and not a 'wolf'. The Maned Wolf inhabits wooded areas and is solitary and nocturnal.

Description: Looks a little like a fox on stilts. A short body with incredibly long legs. The tail is also long and fox-like. Large erect ears, foxy head and shaggy reddish coat.

GIANT ARMADILLO
(*Priodontes giganteus*) ◯

Order: *Edentata*   Family: *Dasypodidae*

Although the Latin name for this Order implies that these animals are all toothless—the description is accurate in the case of the Ant-eaters only. The Giant Armadillo is one of 21 surviving species and it is the largest of all. These strange animals live deep in the forests of South America from south-eastern Venezuela up to north-eastern Argentina.

Description: Upper surface of body, including head, legs and tail, covered with bands of bony plating. The hind legs are exceptionally thick with short, blunt nails on each of the 5 toes. The fore-legs are adapted for digging and four of the toes have long curved claws. The other toe is equipped with a massive claw, the largest claws possessed by any animal in the world today, which is used to rip open ant-hills. Will sometimes walk on hind legs only. Long, sticky tubular tongue.

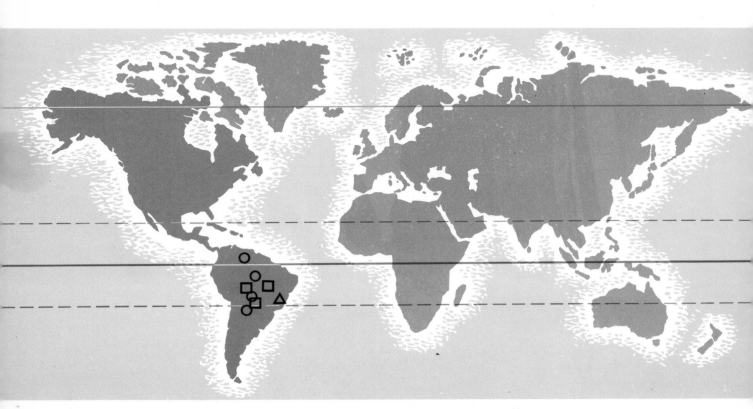

GOLDEN LION MARMOSET
(*Leontideus rosalia*).
Length of body about 8 inches (0·20 metres).
Tail 7 inches (0·17 metres). *Food:* Insects,
seeds, shoots, fruit.

MANED WOLF
(*Chrysocyon brachyurus*).
Length of body 51 inches (1·29 metres).
Tail 16 inches (0·40 metres). Shoulder height
about 2½ ft. (0·76 metres). *Food:* Mammals
from rodents up to sheep, also insects and
birds.

GIANT ARMADILLO
(*Priodontes giganteus*).
Length of body 4 ft. (1·22
metres). Tail 1½ ft. (0·46
metres). Weight up to 100
lbs. (45 kgs.). *Food:* Insects.

CHINCHILLA
(*Chinchilla laniger*) ▲

Order: *Rodentia*   Family: *Chinchillidae*

This attractive squirrel-like rodent lives on the rocky mountain sides of the Andes from Chile to Bolivia. It is now extremely rare in the wild due to over-hunting at the beginning of this Century. Numbers seem to be increasing in some places, thanks to strict protection.

Description: Squirrel-like in appearance, with soft, silky, silvery grey fur and long bushy tail. The Chinchilla has large eyes and delicate cup-like ears. Lives in colonies in burrows. Nocturnal or semi-nocturnal.

VICUNA
(*Vicugna vicugna*) ■

Order: *Artiodactyla*   Family: *Camelidae*

There are four members of the Camel family living in South America and of these, the Vicuna is the smallest. This animal lives in the Andes (in Peru, Argentina, Chile and Bolivia) and inhabits the dry rolling grasslands and plains. Moves in small herds of up to 12 individuals.

Description: Resembles a very slender Llama and is pale fawn in colour. Coat of medium length hair. Short tail. Camel-like head. Lower incisor teeth keep growing (like those of rodents). Will spit evil smelling regurgitated food when angry and will also bite.

PAMPAS DEER
(*Ozotoceros bezoarticus*) ○

Order: *Artiodactyla*   Family: *Cervidae*

This small reddish deer lives on the savannahs and pampas of Brazil, Paraguay, Uruguay, and the Argentine. It is now stated to be the most seriously endangered species of deer in South America, brought to the brink of extinction by over-hunting, the destruction of its natural habitat and the spread of diseases from domestic cattle.

Description: Upper parts and limbs are reddish brown, the face a little darker. Underparts whitish. The tail is dark brown above and white below. Antlers are small, with 3 tines. Bucks have glands on the rear hoofs that give off an unpleasant odour detectable at distances of up to a mile. Can run fast and make jumps of up to 9 ft. (2.74 metres) high.

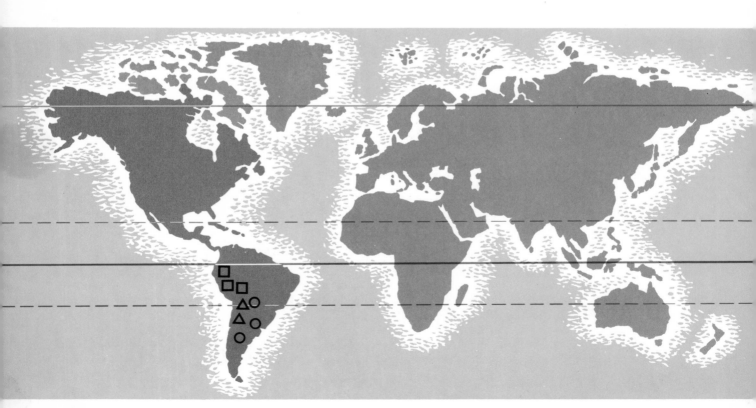

CHINCHILLA
(*Chinchilla laniger*).
Length of head and body 10 inches (0·25 metres).
Tail 10 inches (0·25 metres). *Food:* Grasses, roots
and mosses.

VICUNA
(*Vicugna vicugna*).
Shoulder height 2½ ft. (0·76 metres). Weight 100 lbs. (45 kgs.).
Food: Grasses and herbs.

PAMPAS DEER
(*Ozotoceros bezoarticus*).
Shoulder height about 28 inches (0·71 metres). *Food:* Grasses and herbs.

CUBAN SOLENODON
(*Atopogale cubana*) △

Order: *Insectivora* Family: *Solenodontidae*

Only two species of Solenodon remain in the world today; the Cuban and the Hispaniolan (or Haitian). The Cuban species is right on the verge of extinction with perhaps fewer than 50 left in the wild.

Description: Rat-like in general appearance, with a very long tubular snout bristling with sensitive whiskers. The tail is scaly—but with short, stiff hairs between scales. Coarse, shaggy coat, reddish-brown in colour. Powerful claws on front and hind feet. Musk glands in armpit and groin. Nocturnal.

CUVIER'S HUTIA
(*Plagiodontia aedium*) ▢

Order: *Rodentia* Family: *Capromyidae*

Hutias are primitive rodents which, like the Solenodons, are really living fossils. The genus Plagiodontia now consists of two surviving species Cuvier's Hutia and the Dominican Hutia. Cuvier's Hutia was thought to be extinct until 1947, when it was re-discovered. It is now known to exist in Haiti and the Dominican Republic. The species is almost certainly close to becoming extinct.

Description: A stoutly built blunt-nosed rodent. It has a thick coat consisting of silky hairs, grey with tawny tips mixed with longer black hairs. The underparts are paler in colour and it has long whiskers.

CENTRAL AMERICAN TAPIR
(*Tapirus bairdii*) ○

Order: *Perissodactyla* Family: *Tapiridae*

Also known as Baird's Tapir, the Central American Tapir is the largest of the four surviving species left in the world today. There are three species in Central and South America and one species in Malaya. The Central American Tapir inhabits damp tropical forests in suitable parts of Mexico, Ecuador, Panama and Colombia.

Description: A large, heavy-bodied animal about the size of a pony. The tail is short and stubby. There are four toes on the front feet and three toes on the hind feet. The coat colour is darkish-brown, but it has whitish shading on parts of the face, throat and chest.

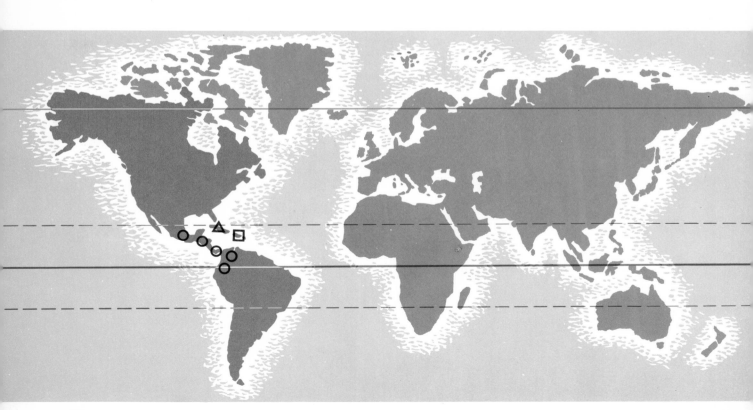

CUBAN SOLENODON
(*Atopogale cubana*).
Length of head and body 1 ft. (0·30 metres). Tail 8 inches (0·20 metres). *Food:* Not strictly insectivorous—and will take small mammals, lizards and carrion in addition to insects.

CUVIER'S HUTIA
(*Plagiodontia aedium*).
Length of head and body about 12 inches (0·30 metres). Tail 5 inches (0·12 metres). *Food:* Leaves, fruit and roots.

CENTRAL AMERICAN TAPIR
(*Tapirus baïrdii*).
Length of head and body about 8 ft. (2·44 metres). Shoulder height 3½ ft. (1·07 metres). Weight 500 lbs. (227 kgs.). *Food:* Grass, leaves and other vegetation.

BLACK-FOOTED FERRET
(*Mustela nigripes*) △

Order: *Carnivora* Family: *Mustelidae*

This large weasel may be the rarest wild mammal in the United States, and probably only a few pairs or individuals still survive within this ferret's range of distribution (across the North American prairies to the east of the Rocky Mountains, from northern Montana and Alberta to central New Mexico).

Description: Large, weasel-like body. The coat is yellowish-buff. Face, throat and belly almost white, with a black mask across its eyes. The legs, feet and tip of tail are black or blackish-brown.

PUMA
(*Felis concolor*) ▫

Order: *Carnivora* Family: *Felidae*

The Puma can still claim a wide range of distribution in spite of the steady decrease in numbers that it has suffered. The Puma may still be found from north-west Canada to the Strait of Magellan, and its habitats vary from mountain tops to desert and jungle. Owing to the fact that the Puma preys on sheep, cattle and horses from time to time it is constantly under attack by man.

Description: A very large, long-tailed Cat with smallish head and reddish-tawny coloured coat. The winter coat is often grey-brown in the colder parts of its range. The young are dull tawny coloured, with profuse black spots.

PRONGHORN
(*Antilocapra americana*) ○

Order: *Artiodactyla* Family: *Antilocapridae*

The Pronghorn is not a true antelope, it is the sole surviving member of a large and ancient family which probably reached its peak some 20 million years ago. It inhabits scrub and desert grasslands in the west and mid-west U.S.A.

Description: Slender antilopine animal, reddish-buff above and whitish below. Black markings on neck, muzzle and in the patch on the throat. Prominent white patches on cheeks, throat and flanks. Upright horns with single forward-projecting branch. The feet are sheep-like and the eyes large.

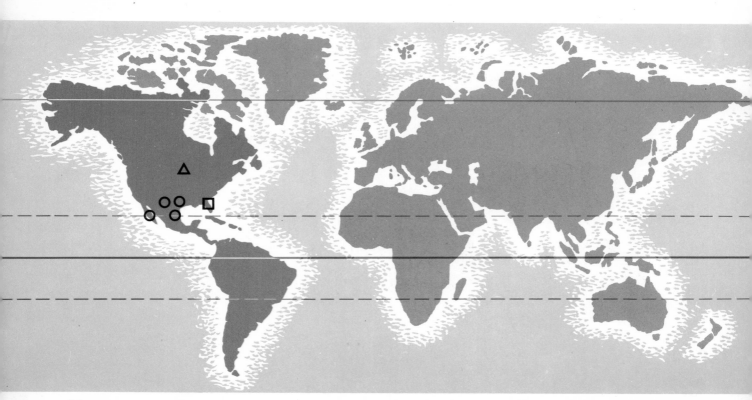

BLACK-FOOTED FERRET
(*Mustela nigripes*).
Length of head and body 20 inches (0·50 metres). Tail 5 inches (0·12 metres). Weight about 1½ lbs. (0·68 kgs.). Females smaller. *Food:* Rodents, including prairie dogs.

PUMA
(*Felis concolor*).
Length of head and body about 7 ft. (2·13 metres). Tail 2½ ft. (0·76 metres). Shoulder height about 28 inches (0·71 metres). Weight around 230 lbs. (104 kgs.). Life span 20 years. *Food:* Deer and a variety of other hoofed and smaller animals.

PRONGHORN
(*Antilocapra americana*).
Length of head and body 4½ ft. (1·37 metres). Tail 4 inches (9·8 cm.). Shoulder height 3½ ft. (1·07 metres). Weight up to 140 lbs. (63 kgs.). *Food:* Vegetation—browses and grazes.

ASIA

The world's largest continent takes its name from the ancient Assyrian name 'asu' or 'land of the East'. Asia covers an area of some 18,500,000 square miles. The coastline measures some 43,000 miles in length and it is washed by numerous seas. The longest river is the Yenisey—which (with the rivers Angara and Selonga included) has a length of 3,690 miles and flows from its source in Mongolia to the Kara Sea (N. Russia). The highest mountain is Mt. Everest with a height of 29,028 ft. (8,838 metres). Asia has a great variety of terrain ranging from the hot deserts of Arabia and central Asia to the frozen tundras of northern Siberia and the steaming jungles and forests of south-east Asia. There are great mountain ranges and inland seas, all kinds of vegetation—and a wonderful profusion of wildlife. Like all the other continents, Asia has its share of animals in danger—and many of these species are very seriously threatened indeed.

SIBERIAN TIGER
(*Panthera tigris altaica*) △

Order: *Carnivora*   Family: *Felidae*

Also known as the Manchurian or Amur Tiger, the Siberian Tiger is the largest of all tigers and the biggest member of the Cat family. This nomadic tiger once ranged widely across Siberia, Mongolia and Manchuria to Korea—but now its range and numbers have been drastically reduced. Only about 250 Siberian Tigers survive in the wild state at the present time—and over 100 of these live in the U.S.S.R. where they are fully and carefully protected. Like the Snow Leopard this tiger is a mountain dweller and it has a long, thick coat to provide protection against the cold. This magnificent cat has always been heavily hunted both by sportsmen and by natives who believe that parts of the tiger are endowed with special magical or medicinal powers! In recent years there can be no doubt that hunting has been responsible for the drastic decline of this species. Solitary by nature, the tiger usually hunts at night—moving stealthily and noiselessly when stalking its prey. The Siberian tiger is also known to hunt during daylight hours. Other tigers in danger include the Bengal tiger, Chinese tiger, Sumatran tiger, Caspian tiger, Javan tiger and Bali tiger.

Description: The coat is palish-fawn in colour with stripes of dark chocolate to black. The tiger has no mane or tail-tuft. The coat is long and thick, and an extra-thick winter coat is grown in the colder months. The underparts are whitish in colour. Largely nocturnal.

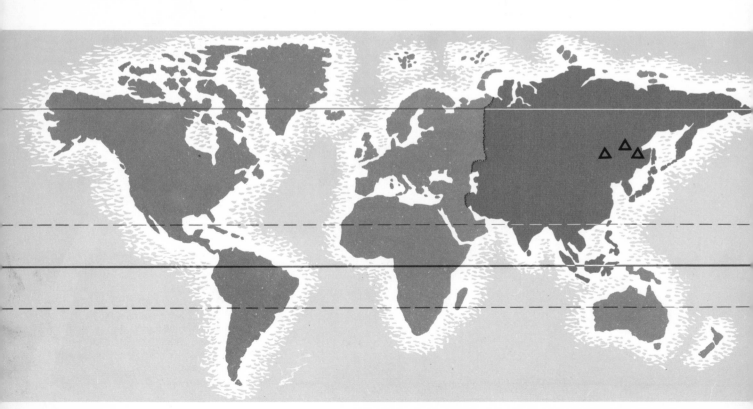

SIBERIAN TIGER
(*Panthera tigris altaica*).

A very large and powerful cat. Head and body 6 ft. (1·83 metres) in length. Tail 3 ft. (0·91 metres). Shoulder height: 38 inches (0·96 metres). Weight: Up to 600 lbs. (272 kgs.). Females slightly smaller. *Food:* Herbivores, large and small domestic cattle, carrion.

SNOW LEOPARD
(*Panthera uncia*) △

Order: *Carnivora*
Family: *Felidae*

Although closely related to the ordinary leopards of Africa and Asia, the beautiful Snow Leopard (or Ounce) is a quite separate species. The long, thick coat provides this cat with much needed protection from the cold conditions prevailing high in the Himalayas and the Altai Mountains of Central Asia where the Snow Leopard lives. During the summer months these leopards can be found living at heights of up to 13,000 ft. (3,962 metres) above sea-level. In the winter they move down to lower levels, following the grazing activities of the animals upon which they feed. They hunt by day or at night. Not yet on danger list—but rare.

Description: A strong and agile cat. It has a long and thickly furred tail. The coat is a pale grey-brown in colour with black rosette markings. Its fur is much thicker and longer than in ordinary leopards, being about 2 inches (4.9 cm.) thick on the back and up to 4 inches (9.8 cm.) thick on the underparts.

ASIATIC LION
(*Panthera leo persica*) ☐

Order: *Carnivora*
Family: *Felidae*

Although the Asian (or Indian) Lion is distinguished by the sub-specific name 'persica', it belongs to the same species as its African counterpart. Contrary to popular supposition it is difficult to find any major differences between the two. The Asiatic Lion may have a rather scantier mane and longer tail tassel than the African Lion, but even in Africa the individual specimens tend to vary considerably. The Asian Lion is now confined to the Gir Forest Reserve, north-west of Bombay. Careful protection brought the numbers up to 290 in 1955. After this there was a decline to 250 in 1961. Further fluctuations occurred until 1967 when it was estimated that the population numbered 261. A survey carried out recently revealed that the number had dropped sharply to 116 specimens, and so the Asiatic Lion is now seriously threatened.

Description: Coat tawny. Mane variable, tawny to black, usually well-developed but may be thin. They live in groups or 'prides' of about ten. Speed up to 50 m.p.h. Inhabit open scrub country, woodland and forest.

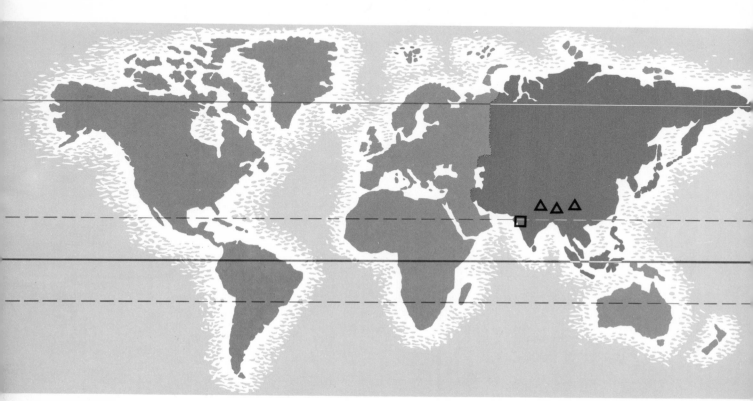

SNOW LEOPARD
(*Panthera uncia*).
Head and body about 4 ft. (1·22 metres) in length. Tail 3 ft. (0·91 metres) long and thickly furred. *Food:* Musk deer, wild sheep, wild goats, bharal and wild boar. Length of life: About 20 years.

ASIATIC LION
(*Panthera leo persica*).
Males attain a length of about 5 ft. (1·52 metres). Tail 3 ft. (0·91 metres). Shoulder height 3 ft. (0·91 metres). Weight 400 lbs. (181 kgs.). Females rather smaller. Length of life: Up to 30 years. *Food:* Nilgai, chital, sambar and wild pigs. Will also take domestic animals.

ORANG-UTAN
(*Pongo pygmaeus*) △

Order: *Primates* Family: *Pongidae*

Living in the low-lying forests of Sumatra and Borneo, the Orang-utan is the only representative of the great apes living outside Africa—and it is certainly the most seriously threatened. Fossil remains show that the orang-utan once enjoyed a much wider distribution than the surviving population of today. Remains of this species have been unearthed in places as far apart as the Siwalik Hills in India and the Chinese provinces of Kwangsi, Kwangtung and Yunnan. The Orang is almost completely arboreal, its extra-long arms being specially adapted for swinging from branch to branch as it moves slowly and lethargically through the dense forest in search of food. Although normally slow-moving, these apes can travel quite quickly when they have to. With an arm-span of around 7 feet (2.13 metres) even their normal unhurried pace carries them along faster than a walking man. At night, the orang-utan builds a 'sleeping platform' of branches in the fork of a tree. Although the platform is only a temporary nest or shelter it may be used by one animal over a period of several nights. Old male orangs are usually solitary by nature but otherwise these apes move in small family groups. When they do come down to ground level they use their long arms as 'crutches' to swing the legs and body along, taking the weight on the knuckles and not the palm of the hand. Young orangs, with their flesh-coloured faces, can look remarkably human and this is probably how these apes came by their name, which in Malay means 'old man of the woods'.

Owing to the demand for orang-utans made by zoos and collectors all over the world, thousands of these apes have been captured or killed in recent years in spite of protection. Normally shy and inoffensive, the orang is not difficult to capture or kill—and it shows little fear of humans. To capture a young orang it is usually necessary to shoot the mother. Often enough the baby dies from the effects of improper feeding, bad handling, disease or parasites long before it reaches a zoo. This constant traffic in young animals and the shooting or trapping of adults is difficult to check due to the determination of poachers. The steady destruction of habitat is also having quite a considerable effect on the orang-utan as many of its forest haunts have been cleared to make way for human settlements, agriculture or industry. Now there are probably fewer than 5,000 orangs left in the wild, and the number continues to decrease.

Description: Heavily built and tailless. Arms long and powerful, legs short and weak. Long, coarse, reddish-coloured hair. Large nostrils, and small ears.

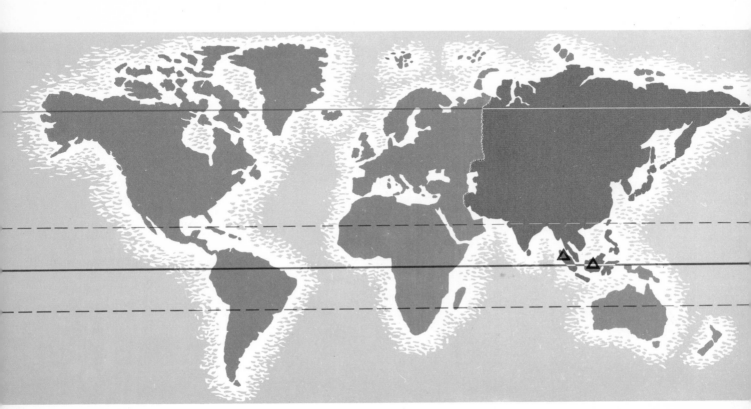

ORANG-UTAN
(*Pongo pygmaeus*).
Height 4½ ft. (1·37 metres). Weight 200 lbs. (91 kgs.). Length of life—up to 26 years. *Food:* Fruit (especially durian).

GIANT PANDA
(*Ailuropoda melanoleuca*) ▲

Order: *Carnivora*   Family: *Procyonidae*

Related to the raccoons, coatis and kinkajous, this popular animal lives in the bamboo forests of Szechwan in south-western China, where it is strictly protected by law. It inhabits an isolated mountainous area and is found at heights ranging from 5,000 (1,524 metres) to 10,000 ft. (3,048 metres) above sea-level.

Description: Bear-like in appearance with a thick, woolly coat. The coat is white, with black legs and shoulders, black ears and black eye patches. The head is large and the teeth are powerful. Diurnal.

CEYLON ELEPHANT
(*Elephas maximus*) ■

Order: *Proboscidea*   Family: *Elephantidae*

This race of Indian Elephant lives in parts of Ceylon and now numbers just over 1,000 specimens left in the wild. Its range has been drastically reduced during the present century due largely to agricultural development.

Description: Smooth trunk (with only one 'lip'). Cows are tuskless. Forehead high and domed. The back convex and ridged. The ears are only one third the size of the ears of the African species. At birth the calves weigh about 200 lbs. (91 kgs) and stand 3 feet (0.91 metres) at the shoulder.

JAPANESE SEA-LION
(*Zalophus californianus japonicus*) ○

Order: *Pinnipedia*   Family: *Otariidae*

Ordinary seals have no visible ear-flaps and have hind limbs that are pretty useless for movement on land. These eared seals however do have small ear-flaps, and they are better adapted for locomotion on land. This close relative of the Californian Sea-lion had its last stronghold around the rocky islet of Take-Shima in the Sea of Japan. Since the occupation of this islet by South Korean troops in the 1950's, the Japanese Sea-lion has disappeared, and may well be extinct.

Description: Although these animals appear to be black when wet; when dry, the coat colour ranges from light buff to dark brown. The adult male has a crest on top of the head. They have large, powerful flippers.

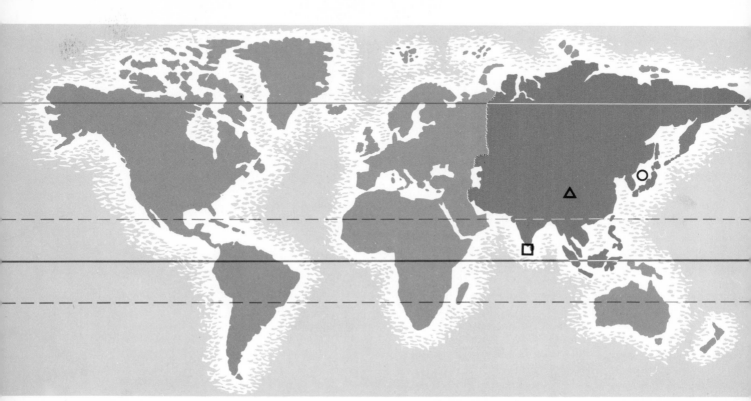

GIANT PANDA
(*Ailuropoda melanoleuca*).
Head and body 6 ft. (1·83 metres)
in length. Tail a mere stump.
Weight 300 lbs. (136 kgs.). *Food:*
Chiefly bamboo shoots. Also be-
lieved to take small mammals and
birds.

CEYLON ELEPHANT
(*Elephas maximus*).
Bulls stand 9 ft. (2·74 metres) at
the shoulder and weigh up to
4 tons (4,063 kgs.). Cows are rather
smaller. Tusks usually short.
Walking speed: 4 m.p.h. Charg-
ing speed: 30 m.p.h. *Food:* Fruit,
grasses, reeds, shoots, bamboo
and leaves, about 400 lbs. (181
kgs.) per day. Drink up to 50
gallons (187 litres) of water daily.

JAPANESE SEA-LION
(*Zalophus californianus japonicus*).
Bulls may attain a length of 7 ft. (2·13
metres) and weigh up to 600 lbs.
(272 kgs.). *Food:* Fish, shellfish, squid
and seabirds.

BACTRIAN CAMEL
(*Camelus bactrianus*) △

Order: *Ungulata* Family: *Camelidae*

Easily distinguished from the Arabian Camel by its two fatty humps, shorter legs and longer, shaggier coat, the Bactrian or Asiatic Camel is a native of the cold deserts of Central Asia. About 400 of these animals still exist in a wild state deep in the Gobi Desert. Other tiny groups may also exist in parts of China.

Description: The wild Bactrian Camel is less heavily built and rather less hairy than the domestic form. The humps too, are smaller. Coat colour varies from light grey to dark brown. Feet are well adapted for movement in rough hilly country. Will live at heights of over 11,000 ft. (3,353 metres) above sea-level during the summer and then move down to the desert again for the winter.

PRZEWALSKI'S HORSE
(*Equus przewalskii*) □

Order: *Perissodactyla* Family: *Equidae*

Also known as the Mongolian Wild Horse, this pony-sized animal now numbers fewer than 40 specimens in the wild state and is the last surviving species of wild horse in the world. The few remaining Przewalski's Horse live on the barren plains of the Altai Mountains and in the west of Mongolia.

Description: Differs from the domestic horse in having a short, erect mane, heavy head, small ears and a low-slung tail. Coat colour is pale brown. A long winter coat is grown. The mane and tail are black. The muzzle white.

ASIATIC WILD ASS
(*Equus hemionus onager*) ○

Order: *Perissodactyla* Family: *Equidae*

The five races of Asiatic Wild Ass are all threatened with extinction. Our picture shows the Persian Wild Ass (or Onager). The other three races are the Indian Wild Ass (or Ghor-khar), the Tibetan Wild Ass (or Kiang) and the Mongolian Wild Ass (or Kulan). The Asian asses differ from the African species by their shorter, narrower ears and their colour. African asses are greyish, the Asian asses are yellowish. They all live in herds.

Description: (Onager). Fawny-white upper surface with blackish dorsal stripe. Tail-tuft and mane also black. Black tips to their ears. Underparts white.

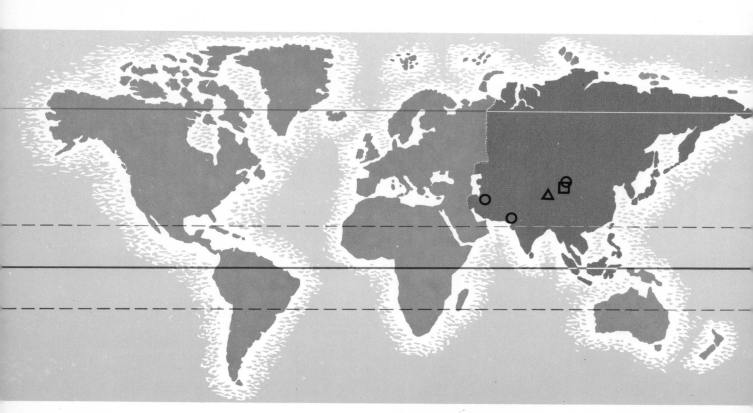

BACTRIAN CAMEL
(*Camelus bactrianus*).
Shoulder height 5 ft. (1·52 metres) or more.
Food: Dry, prickly scrub and similar desert vegetation.

ASIATIC WILD ASS
(*Equus hemionus*).
Our illustration shows the Persian Wild or Onager (*Equus hemionus onager*). This animal stands about 3½ ft. (1·07 metres) at the shoulder. *Food:* Grasses.

PRZEWALSKI'S HORSE
(*Equus przewalskii*).
Stocky and pony-like. Shoulder height 4½ ft. (1·37 metres). *Food:* Grasses.

GREAT INDIAN RHINOCEROS
(*Rhinoceros unicornis*) △

Order: *Perissodactyla*   Family: *Rhinocerotidae*

Five species of rhinoceros survive in the world today, three of them living in Asia and the other two in Africa. Living on the grassy plains of Nepal, Bengal and Assam, the Great Indian Rhinoceros is the largest of the three Asian species—and the most numerous. At the present time about 750 of these animals exist in the wild—but the number is constantly being threatened by the activities of very determined poaching gangs who raid the rhino sanctuaries in spite of the efforts made to check them. In one year (1966) poachers killed 13 rhinos in one sanctuary alone (the Kaziranga Sanctuary in Assam). The name rhinoceros means 'horned-nose', and this is an apt description of the prehistoric looking rhino. Unfortunately for all the five species of rhinoceros in the world today it is that 'horned-nose' which marks them down as targets for poaching gangs. The horn itself is simply a mass of tightly compressed fibres and is of no real value (unlike the ivory tusks of the elephant). However, in parts of the Far East, powdered rhino horn is considered to have great magical powers—and so a very high price is paid for the horn. The price paid to the poachers is high enough to make it worth risking severe penalties if they are caught raiding the sanctuaries

—and so the hunting goes on. Even today this superstition about the power of rhino horn still prevails, and so the rhinoceroses in Asia and Africa will need full protection if they are going to survive.

In spite of its rather formidable appearance, the Great Indian Rhinoceros is normally shy and harmless unless provoked. If provoked, of course, the animal can be dangerous and although the horn in this species is short and blunt, the Great Indian rhino has two sharp-edged tusks in the lower jaw which can be used very effectively as slashing weapons.

Of the Asian species the Great Indian and the Javan rhino have only one horn while the Sumatran rhino has two.

Description: Blackish-grey in colour. The 'studded' skin is heavily folded in front of and behind the shoulder and in front of the thigh. This gives the animal an 'armour-plated' appearance. There are also folds of skin around the neck and even the tail is set in a deep slot in the skin-folds. The legs are short and stout and there are three toes on each of the feet. The horn is short and blunt. Hair is visible only on ear-fringe and tail-tip.

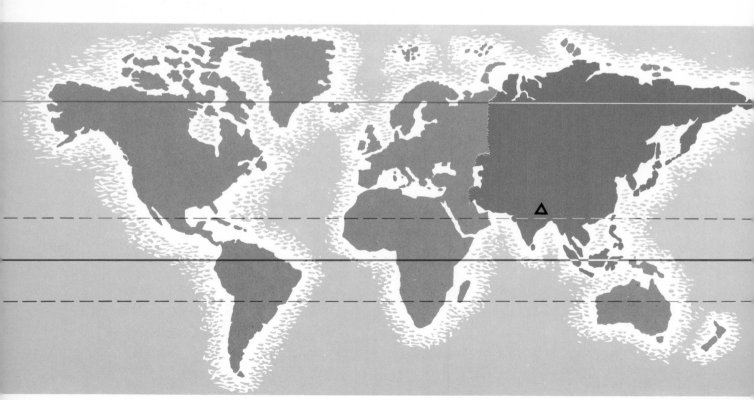

GREAT INDIAN RHINOCEROS
(*Rhinoceros unicornis*).
Length 14 ft. (4·22 metres). Tail 2 ft. (0·61 metres). Shoulder height 6 ft. (1·83 metres). Weight about 2 tons (2,032 kgs.). Length of horn up to 2 ft. (0·61 metres). *Food:* Grasses and other vegetation.

JAVAN RHINOCEROS
(Rhinoceros sondaicus) △

Order: *Perissodactyla*   Family: *Rhinocerotidae*

Probably the rarest large animal in the world today, the Javan species is certainly the rarest of all the rhinos and only about 25 are believed to exist in the wild. They inhabit thick jungle and marshland, but may also be found in forested mountain country. The male carries a short blunt horn. Females are usually hornless. All three of the Asiatic species of rhino used to be found in India and the Javan Rhinoceros was fairly common there until about 1900. The decline of this animal has been brought about by the great increase in human population and the subsequent spread of agriculture and settlement. This coupled with the heavy hunting for the horn has reduced the species to its present state. In an effort to protect and preserve the surviving population the Udjung Kulon Reserve was established on the western tip of Java some years ago. Thanks to this protection the species seems to stand some chance of survival.

Description: Skin blackish-grey and hairless. The skin is heavily folded as in the Indian species, but with a mosaic of wrinkles instead of the 'studding' found on the skin of the Indian Rhinoceros.

SUMATRAN RHINOCEROS
(Didermocerus sumatrensis) ☐

Order: *Perissodactyla*   Family: *Rhinocerotidae*

Although more widespread than the Javan species, the small, hairy two-horned Sumatran Rhinoceros is also very rare—and only about 100–170 remain in the wild. This rhinoceros is easily distinguished from the other two Asian species by its smaller size, hairy skin and its two horns. The Sumatran Rhinoceros is, in fact, the smallest and hairiest of all the five species of rhino. The habits of the Sumatran rhino are similar to those of its Javan cousin. Both favour hill forests and both live close to water or marshland. Although the species is protected by law throughout most of its range—it is extremely difficult to enforce this protection because of the wandering habits of the animal. So little is known about the life and habits of this species that a thorough survey will be necessary before plans can be prepared for the future preservation of the Sumatran rhino.

Description: The skin is less well folded, with just one complete fold behind the shoulder and is smoother than in the other two Asian species. It is covered with a thin layer of coarse hair. Food: Leaves and twigs.

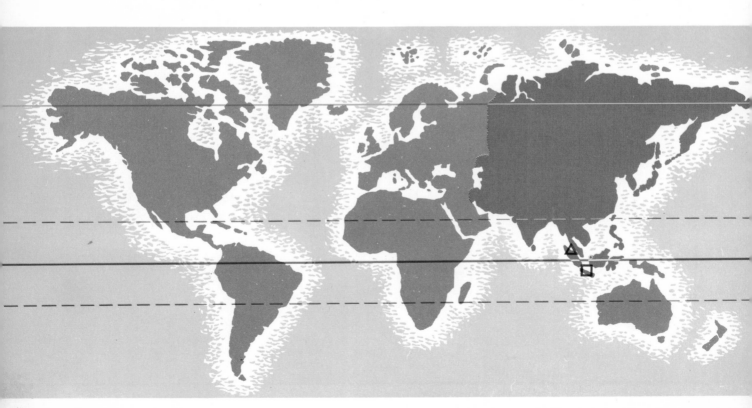

JAVAN RHINOCEROS
(*Rhinoceros sondaicus*).
Adults may grow to 10 ft. (3·05 metres) in length
and stand 5½ ft. (1·68 metres) at shoulder. Weight
about 1 ton (1,016 kgs.). Length of horn—up to
5 inches (0·12 metres). *Food:* Leaves and twigs.

SUMATRAN RHINOCEROS
(*Didermocerus sumatrensis*).
Length 8 ft. (2·44 metres). Shoulder height 4½ ft.
(1·37 metres). Weight up to 1 ton (1,016 kgs.).
Front horn about 20 inches (0·51 metres). Rear
horn about 6 inches (0·15 metres). *Food:* Leaves
and twigs.

THAI BROW-ANTLERED DEER
(*Cervus eldi siamensis*) △

Order: *Artiodactyla* Family: *Cervidae*

There are three races of this south-east Asian deer but only two of these are in danger of extinction. Our picture shows the Thailand Brow-Antlered Deer which is found in Thailand and Viet-Nam. This race is fast disappearing and may not survive much longer unless carefully protected. It is not known just how many of the Thai race still survive.

Description: Related to the Swamp Deer they inhabit similar marshy regions. In summer the coat is reddish-brown with white below. In winter dark brown with lighter coloured underparts. Coat spotted along middle line of back and occasionally on the sides.

FORMOSAN SIKA
(*Cervus nippon taiouanus*) □

Order: *Artiodactyla* Family: *Cervidae*

Often referred to as the most handsome of all deer the Formosan Sika may already be extinct in the wild. About 300 specimens exist in zoos and private collections in various parts of the world. They live in mountain forests.

Description: Summer coat colour light chestnut, with large white spots and a deep reddish tinge on the hinder part of the neck. The winter coat is longer, darker and less spotted. There is a strongly marked dorsal stripe.

PERE DAVID'S DEER
(*Elaphurus davidianus*) ○

Order: *Artiodactyla* Family: *Cervidae*

Originally widespread across the plains of north-east China, the species was first observed in 1865 by the eminent French missionary and naturalist Abbe Armand David. Although extinct in the wild, there are now rather more than 400 of these deer in zoos and private collections.

Description: Reddish-tawny coat with white underparts. There is a whitish ring around the eye. Tail long and tufted. The antlers are quite long and they are unusual in that the front prong is forked and the rear prong is usually straight. Antlers may be shed twice in one year.

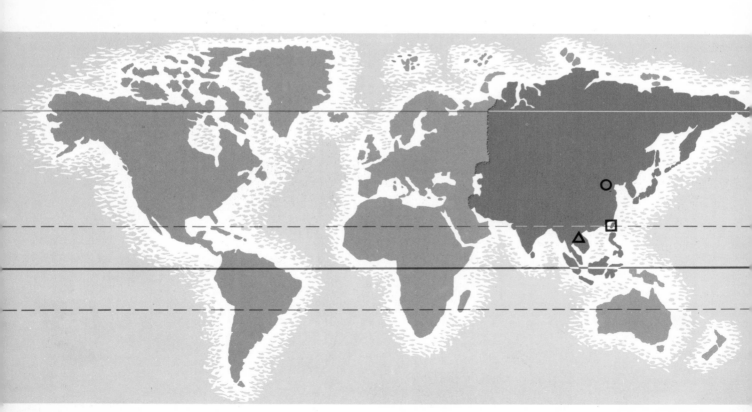

THAI BROW-ANTLERED DEER
(*Cervus eldi siamensis*).
Shoulder height about 4 ft. (1·22 metres).
Weight up to 250 lbs. (113 kgs.). Good
antlers up to 40 inches (1·01 metres) in
length. *Food:* Grasses and leaves.

FORMOSAN SIKA
(*Cervus nippon taiouanus*).
Shoulder height about 3 ft. (0·91 metres).
Antlers about 18 inches (0·46 metres).
Food: Grasses and leaves.

PERE DAVID'S DEER
(*Elaphurus davidianus*).
Shoulder height 45 inches (1·14 metres).
Antlers 30-35 inches (0·76-0·88 metres) in
length. Young are boldly spotted. *Food:*
Grasses, leaves and water plants.

ANOA
(*Anoa depressicornis*) △

Order: *Artiodactyla* Family: *Bovidae*

The Anoa (Dwarf Buffalo or Wood-Ox) is the smallest wild member of the family Bovidae. This shy animal lives in remote wooded mountains on the island of Celebes where it is hunted for its meat. It usually keeps close to water. Little is known about the Anoa's habits in the wild. There are three recognised races of Anoa on the island.

Description: Small, but sturdy in build, with short legs and plump body. Horns short and straight about 12 inches (30.48 cm) in length. The coat is very thin and is dark-brown in colour. Young animals are yellowish-buff coloured and quite woolly.

YAK
(*Bos grunniens*) ☐

Order: *Artiodactyla* Family: *Bovidae*

This species of wild cattle inhabits the uplands of Tibet and may be found at altitudes of up to 20,000 ft. (6,096 metres). Even in the winter they can thrive at heights of up to 15,000 ft. (4,572 metres)! The Yak has been domesticated in Tibet for centuries and it is still the typical beast of burden in this part of Central Asia. It is very sure footed.

Description: Heavily built and powerful. Short legs with large rounded feet. Large wide-spreading horns. Coat smooth but very long on lower parts of animal. Long fringe of hair hangs from its flanks down to its ankles. Coat colour blackish-brown with whitish muzzle.

MARKHOR
(*Capra falconeri*) ○

Order: *Artiodactyla* Family: *Bovidae*

Largest of all the wild goats, the Markhor lives in the Himalayas from Afghanistan to Kashmir. Four races of this goat are recognised. An extremely agile animal, the Markhor inhabits precipitous forested mountain country. Among their natural enemies are Snow Leopards and Wild Dog packs. Hunting by man and the transmission of diseases by domestic cattle have helped to reduce this goat to its present precarious population level.

Description: A heavily built goat with long, spirally twisted horns. Old males have a long beard. Coat is short in the summer but long in the winter. In summer the coat is reddish-brown, in winter grey.

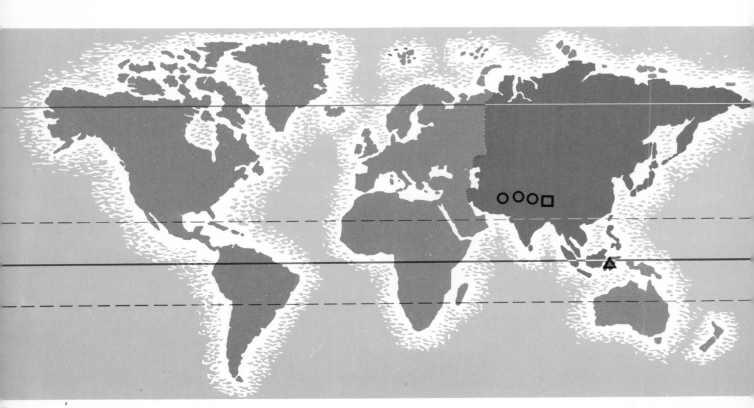

ANOA
(*Anoa depressicornis*).
Body length about 5½ ft. (1·68 metres).
Shoulder height 40 inches (1·01 metres).
Tail 10-12 inches (0·26-0·30 metres).
Horns short and straight—about 12
inches (0·30 metres) in length. *Food:*
Vegetation.

YAK
(*Bos grunniens*).
Shoulder height up to 5½ ft. (1·68 metres). Weight
around 1,200 lbs. (544 kgs.). Horns up to 38
inches (0·96 metres). *Food:* Coarse mountain
grasses. Drink water in summer and eat snow in
winter if water not available.

MARKHOR
(*Capra falconeri*).
Shoulder height 40 inches (1·01 metres).
Weight around 200 lbs. (91 kgs.). Horns
4 ft. (1·22 metres) in length. *Food:*
Grasses. Name means 'snake-eater'—but
there is no evidence to suggest that it
actually does eat them!

AUSTRALASIA

Australia is the world's largest island and smallest continent. It covers an area of 2,974,581 square miles and provides a home for some of the most fascinating animal life to be found in the world.

BRIDLED NAIL-TAIL WALLABY
(*Onychogalea fraenata*) △

Order: *Marsupialia* Family: *Macropodidae*

This hare-sized wallaby is now extremely rare throughout its range in Eastern Australia. None are recorded in captivity.

Description: A most attractive wallaby, shy and timid, takes its name from the head and shoulder markings which suggest a bridle. The tail has a horny tip not unlike a finger nail. The coat is soft and silky.

GREY CUSCUS
(*Phalanger orientalis*) □

Order *Marsupialia* Family: *Phalangeridae*

Found in the tropical rainforests of northern Australia, the Grey Cuscus is becoming increasingly rare.

Cuscuses are the largest members of the phalanger family, the Australian opossums. Two species live in Australia, Grey Cuscus and Spotted Cuscus.

Description: A cat-sized marsupial, with an elongated body and a prehensile, scaly tail. Arboreal and sluggish.

KOALA
(*Phascolarctos cinereus*) ○

Order: *Marsupialia*. Family: *Phalangeridae*

Although not officially classified as an 'animal in danger'—the Koala has been drastically reduced in numbers both by Man and disease since the turn of the century. It is now strictly and carefully protected by law.

Description: Slow moving, tree-dwelling member of the phalanger family. The Koala is tailless with a woolly coat and prominent tufted ears. The coat is ash-grey in colour, tinged with brown on the upper parts and a yellowish-white on the underparts and hind quarters. A highly specialised feeder, eating about 3 lb (1.36 kgs) of leaves daily. The name 'Koala' is Aborigine for 'no drink'. The animal gets its moisture from the leaves. Diurnal.

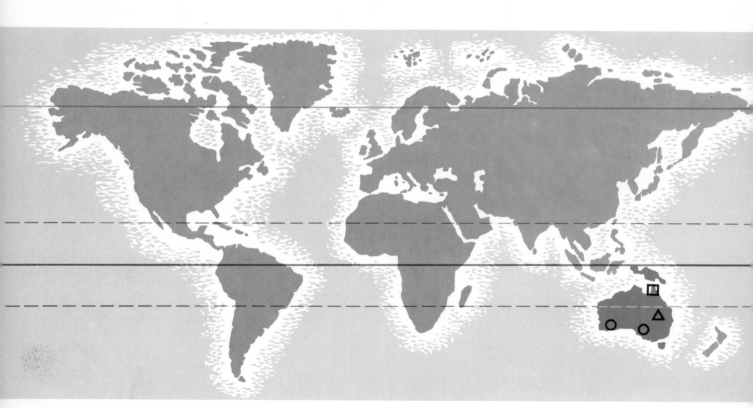

BRIDLED NAIL-TAIL WALLABY
(*Onychogalea fraenata*).
Head and body 22 inches (56 cm.), tail 18 inches (46 cm.). *Food:* Grass, small plants, bulbous roots. Diurnal.

GREY CUSCUS
(*Phalanger orientalis*).
The Grey Cuscus has a thick woolly coat, small ears, big round eyes and half-naked tail. Head and body up to 26 inches (65 cm.). Tail up to 24 inches (61 cm.). *Food:* Mainly leaves and fruit but also known to take small birds and reptiles. Nocturnal.

KOALA
(*Phascolarctos cinereus*).
Bear-like in appearance and about 2 feet (61 cm.) in length. *Food:* Eats the leaves from certain kinds of eucalyptus trees.

THYLACINE
(*Thylacinus cynocephalus*) △

Order: *Marsupialia*   Family: *Dasyuridae*

Also known as the Tasmanian Tiger or Tasmanian Wolf, the Thylacine is the largest known flesh-eating marsupial. Formerly common in Tasmania it is now on the brink of extinction.

Description: The head is dog-like. The rump and tail are not unlike those of the kangaroo. Coat colour greyish-brown with up to 18 dark-chocolate coloured bands extending from saddle to the root of the tail. The female has a backward-facing brood pouch. Nocturnal by nature.

NATIVE CAT
(*Dasyurus quoll*) □

Order: *Marsupialia*   Family: *Dasyuridae*

This cat-sized marsupial is found in S.E. Australia and Tasmania. Although not in any immediate danger of extinction the Native Cat has been drastically reduced in numbers on the Australian mainland and now exists only in isolated areas. Still quite common in Tasmania, it spends the day asleep in small caves or holes, in hollow trees or among rocks.

Description: Olive-grey coat covered with small white spots on head and body. Tail not spotted. Nocturnal.

NUMBAT
(*Myrmecobius fasciatus*) ○

Order: *Marsupialia*   Family: *Dasyuridae*

This banded Ant-eater lives in the open woodlands of S. W. Australia. Unlike other Marsupials the female has no brood pouch. Terrestrial and Diurnal.

Description: Underparts yellowish. It has a slender snout and 4 inches (8.9 cm) long sticky tongue. Powerful claws on front feet and bushy tail.

HAIRY-NOSED WOMBAT
(*Lasiorhinus latifrons*) ◇

Order: *Marsupialia*   Family: *Vombatidae*

Although drastically reduced in numbers during the past few years the Hairy-Nosed Wombat is still fairly common in a few scattered areas in South Australia.

Description: A thickset animal with short and powerful legs. Stump-like tail. Strong digging claws on all four feet. Coat soft, almost silky and grizzled grey in colour. Nocturnal.

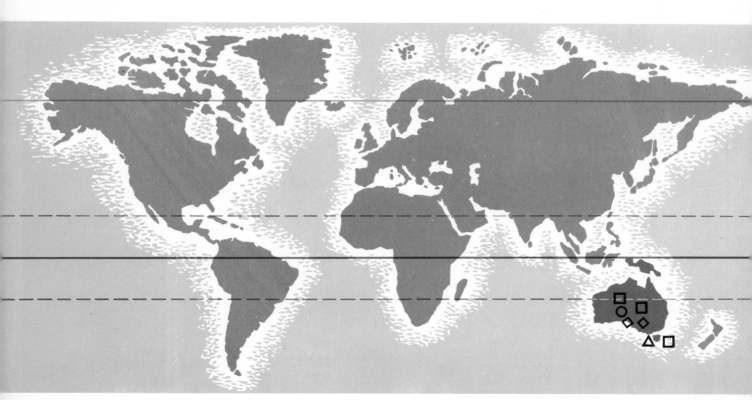

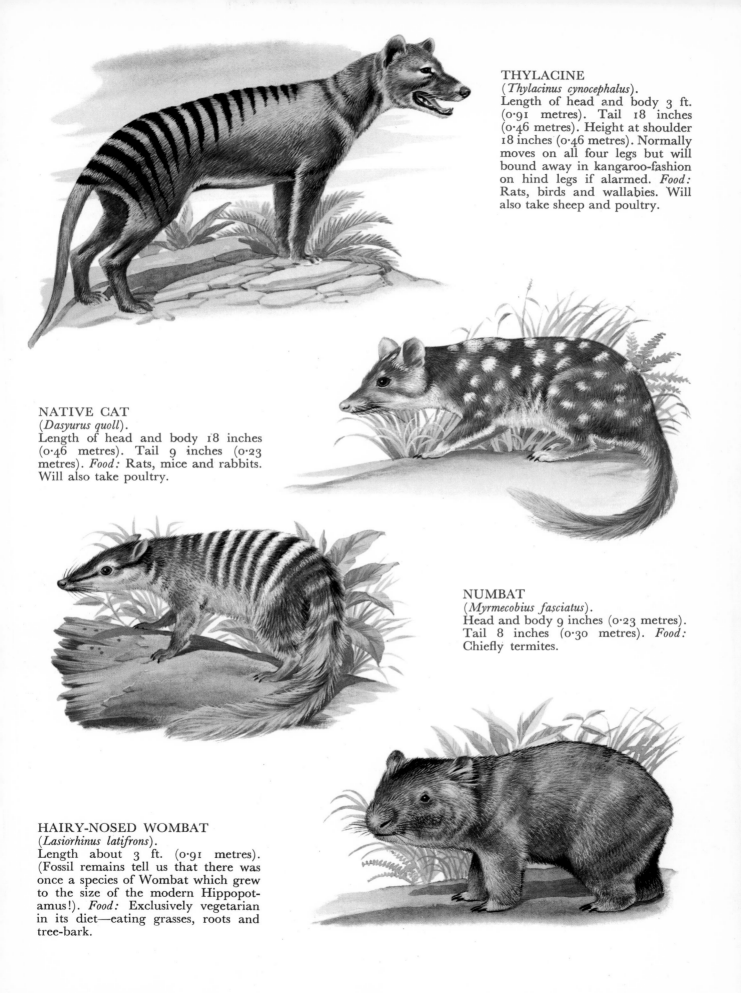

THYLACINE
(*Thylacinus cynocephalus*).
Length of head and body 3 ft. (0·91 metres). Tail 18 inches (0·46 metres). Height at shoulder 18 inches (0·46 metres). Normally moves on all four legs but will bound away in kangaroo-fashion on hind legs if alarmed. *Food:* Rats, birds and wallabies. Will also take sheep and poultry.

NATIVE CAT
(*Dasyurus quoll*).
Length of head and body 18 inches (0·46 metres). Tail 9 inches (0·23 metres). *Food:* Rats, mice and rabbits. Will also take poultry.

NUMBAT
(*Myrmecobius fasciatus*).
Head and body 9 inches (0·23 metres). Tail 8 inches (0·30 metres). *Food:* Chiefly termites.

HAIRY-NOSED WOMBAT
(*Lasiorhinus latifrons*).
Length about 3 ft. (0·91 metres). (Fossil remains tell us that there was once a species of Wombat which grew to the size of the modern Hippopotamus!). *Food:* Exclusively vegetarian in its diet—eating grasses, roots and tree-bark.

EUROPE

Europe is the second smallest of the world's continents after Australia and covers an area of about 4,060,184 square miles. The coastline of the continent measures some 23,550 miles. The longest river is the Volga (U.S.S.R.) with a length of 2,293 miles—and the highest point is usually named as Mont Blanc (France) with a height of 15,770 ft. (4,807 metres). Largely due to the density of the human population, which is constantly increasing, the wild lands and the wildlife of Europe have been steadily pushed out.

SPANISH LYNX
(*Felis lynx pardina*) △

Order: *Carnivora* Family: *Felidae*

Once common throughout the Iberian Peninsula. It is now confined to a few areas in the extreme south of Spain where only a few hundred may now survive.

Description: Medium-sized and cat-like, but with very short tail and tufted ears (common to all members of the lynx tribe). The limbs are long and powerful. Coat short and light rufous in colour marked with black stripes and rows of black spots. Underparts whitish.

EUROPEAN BISON
(*Bison bonasus*) ▢

Order: *Artiodactyla* Family: *Bovidae*

The European Bison or Wisent (German) is now extinct in the wild but about 860 still survive in forest reserves and parks in parts of Europe.

Description: Large, powerful wild cattle with massive shoulders, humped back and smallish head. Hind quarters slope downwards. Eyes small. Fairly shaggy chestnut-brown coat. Both sexes are horned.

COMMON OTTER
(*Lutra lutra*) ◇

Order: *Carnivora* Family: *Mustelidae*

Although still quite widespread throughout Europe, the Otter has been steadily declining in numbers during the past few years in Britain.

Description: A long-bodied *water-weasel* with short legs, webbed feet, flattish head and a long, thick, tapering tail (*or rudder*). The fur is dense and dark brown in colour with paler underparts. Eyes small. Muzzle broad. Ears small and partly hidden.

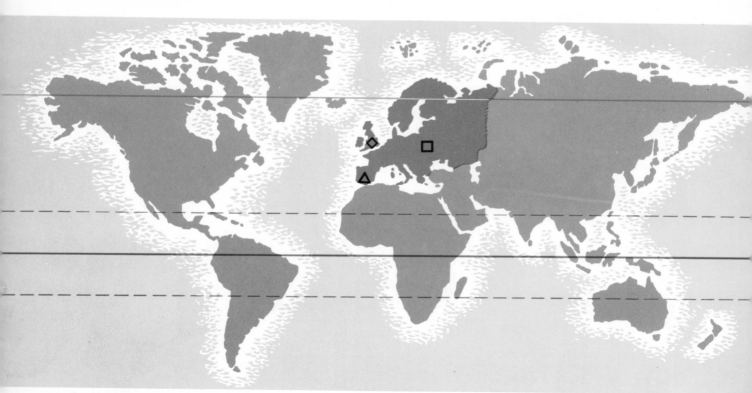

SPANISH LYNX
(*Felis lynx pardina*).
Length of head and body 3 ft. (0·91 metres).
Tail 5 inches (0·13 metres). Shoulder height
2 ft. (0·61 metres). Weight about 25 lbs.
(11 kgs.). *Food:* Hares, rabbits, birds, small
deer.

EUROPEAN BISON
(*Bison bonasus*).
Length of head and body up to 9 ft.
(2·74 metres). Tail 2½ ft. (0·76 metres).
Shoulder height 5 ft. (1·52 metres).
Weight around 1,800 lbs. (816 kgs.).
Food: Grass.

COMMON OTTER
(*Lutra lutra*).
Length of head and body 2½ ft. (0·76 metres).
Tail 1½ ft. (0·46 metres). Weight up to
25 lbs. (11 kgs.). Shoulder height 7 inches
(0·18 metres). Life span up to 15 years.
Food: Chiefly fish—but also mammals,
birds, frogs and some invertebrates.

THE OCEANS

POLAR BEAR
(*Thalarctos maritimus*) △

Order: *Carnivora*
Family: *Ursidae*

The Polar Bear is one of the largest of bears and certainly by far the most carnivorous member of the family.

Living in the Arctic regions of America, Europe and Asia, this bear is on the danger list throughout its range of distribution and its chief predator is Man.

The Polar Bear is a solitary wanderer and great traveller. During the bitter Arctic winter the males tend to move southward while the females build a snow den and spend much of their time under cover. There is no actual hibernation but the females may sleep for days at a time inside the den. During this period the cubs are born. There are usually two cubs at birth and they are about the size of rabbits when born.

Polar Bears are strong swimmers and expert divers.

Description: The coat is white with a yellowish tinge. The head is long and so is the neck. These bears have hairy pads on the soles of their feet which not only help to insulate them against the cold but also provide a good gripping surface on the ice. The tail is nothing more than a stump.

STELLER'S SEA-COW
(*Hydrodamalis stelleri*) ▢

Order: *Sirenia*   Family: *Dugongidae*

The order Sirenia is divided into two groups— Dugongs and Mantaees.

Steller's Sea-Cow, a species of dugong, lived in the Bering Sea north of the Aleutian Islands. Many world authorities claim that this animal has been extinct since 1768, but in 1962 hopes were raised that a small population of survivors was still living. This is now considered to be highly improbable.

The discovery of this sea-cow in 1741 was followed by an absolute massacre during the next twenty years as traders and hunters killed the animals for their meat.

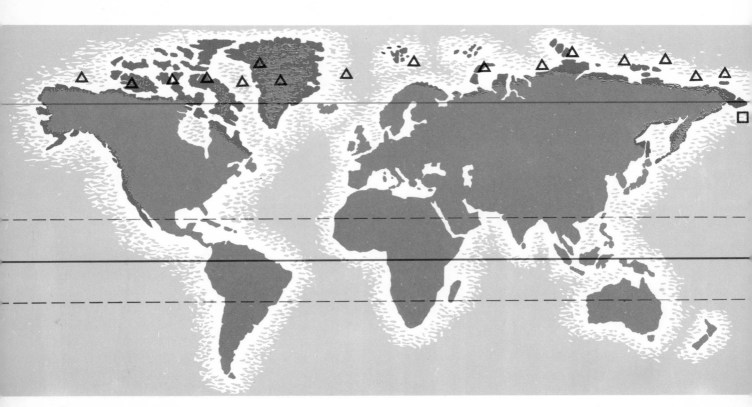

POLAR BEAR
(*Thalarctos maritimus*).
Males may reach a length of 8 ft. (2·44 metres), and weigh up to 1,600 lbs. (725·6 kgs.). *Food:* Chiefly seals plus any other animals he is able to catch. Will also eat grass, moss and lichen.

STELLER'S SEA-COW
(*Hydrodamalis stelleri*).
Toothless aquatic mammals, ungainly, about 24 ft. (7·33 metres) in length and weighing well over a ton (1,015·84 kgs.). *Food:* Underwater vegetation.

MANATEE
(*Trichechus manatus*) △

Order: *Sirenia*   Family: *Trichechidae*

There are three recognised species of this ungainly and rather seal-like creature. The North American Manatee can still be found along the coasts of the Caribbean, the coastal rivers of Central and South America and the Florida Peninsula; the South American Manatee is found in the lower Amazon. The West African Manatee is now very rare and found in lagoons and the lower reaches of West African rivers between Senegal and Angola.

Description: Large, paddle-shaped horizontal tail. Small head and practically no neck. Bulbous upper lips and valvular nostrils. No external ear. No front teeth in adults. Coarse bristles on muzzle. Grey-black in colour, with forelimbs modified into flippers.

SOUTHERN SEA OTTER
(*Enhydra lutris nereis*) □

Order: *Carnivora*   Family: *Mustelidae*

The Sea Otter is entirely marine in its habits and rarely, if ever, comes out on land. Between 800 and 900 of these fascinating 'marine weasels' still survive in the waters of Carmel Bay and off parts of the Monterey Peninsula (California). Where they are fully protected by law.

Description: Long-bodied, with flattened head, small ears, broad snout and small front legs with short webbed feet. Large webbed hind feet, broad and flipper-like. Brownish-black thick glossy fur sprinkled with white-tipped hair. It will swim and float on its back, and also sleep on its back— anchored to seaweed. Whilst floating on its back it will hold a rock against its chest and use it as an anvil for breaking open shellfish.

ATLANTIC WALRUS
(*Odobenus rosmarus*) ○

Order: *Pinnipedia*   Family: *Odobenidae*

It has been estimated that only about 25,000 Atlantic Walruses may still be living in the Arctic waters from the Kara Sea to Hudson Bay. There is just one species of walrus, divided into two races (the other being the now safe Pacific Walrus).

Description: Large and stout, with small head, square muzzle and 2 large tusks (canine teeth) in the upper jaw. Skin wrinkled and folded. Fur short and brownish-grey. Coarse white bristles on snout. Old bulls are often almost hairless. Hind limbs are bent forward when ashore or on ice. Comparatively weak forelimbs. No external ear.

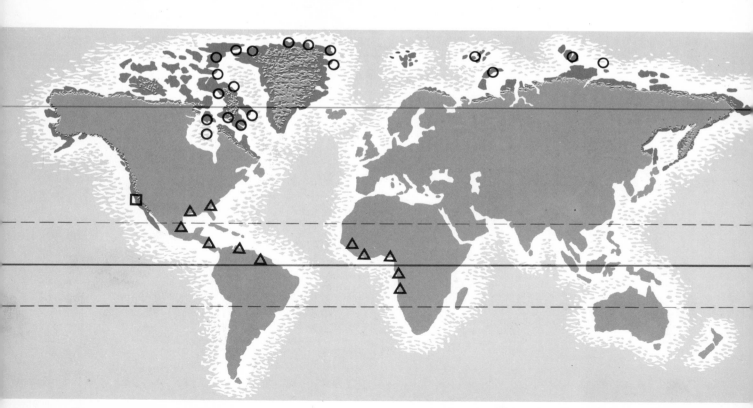

MANATEE

(*Trichechus manatus*).
Length 12 ft. (3·66 metres) or more. Weight 500-1,000 lbs. (227-454 kgs.). *Food:* Seaweed and other marine vegetation up to 100 lbs. (45 kgs.) daily.

SOUTHERN SEA OTTER

(*Enhydra lutris nereis*).
Length of head and body 3 ft. (0·91 metres). Tail 1 ft. (0·30 metres). Weight 80-85 lbs. (36-39 kgs.). Can remain submerged 4-5 minutes. *Food:* Fish, crabs, molluscs, sea-urchins.

ATLANTIC WALRUS

(*Odobenus rosmarus*).
Length about 12 ft. (3·66 metres). Weight up to 3,000 lbs. (1,361 kgs.). Length of tusks over 2 ft. (0·61 metres). Weight of tusk up to 11 lbs. (4·99 kgs.). *Food:* Crustaceans, molluscs, etc.